5th Grade Math Volume 3

© 2013 OnBoard Academics, Inc
Newburyport, MA 01950
800-596-3175
www.onboardacademics.com

ISBN: 978-1494857325

Table of Contents

Decimals

Key Vocabulary

tenth

hundredth

thousandth

expanded notation

Decimal Concept

What does the ? represent. Write it as a decimal.

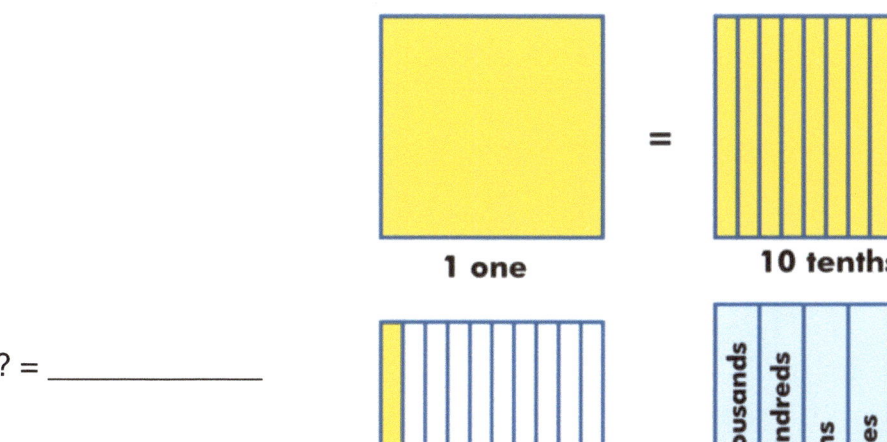

? = _____

Try another problem

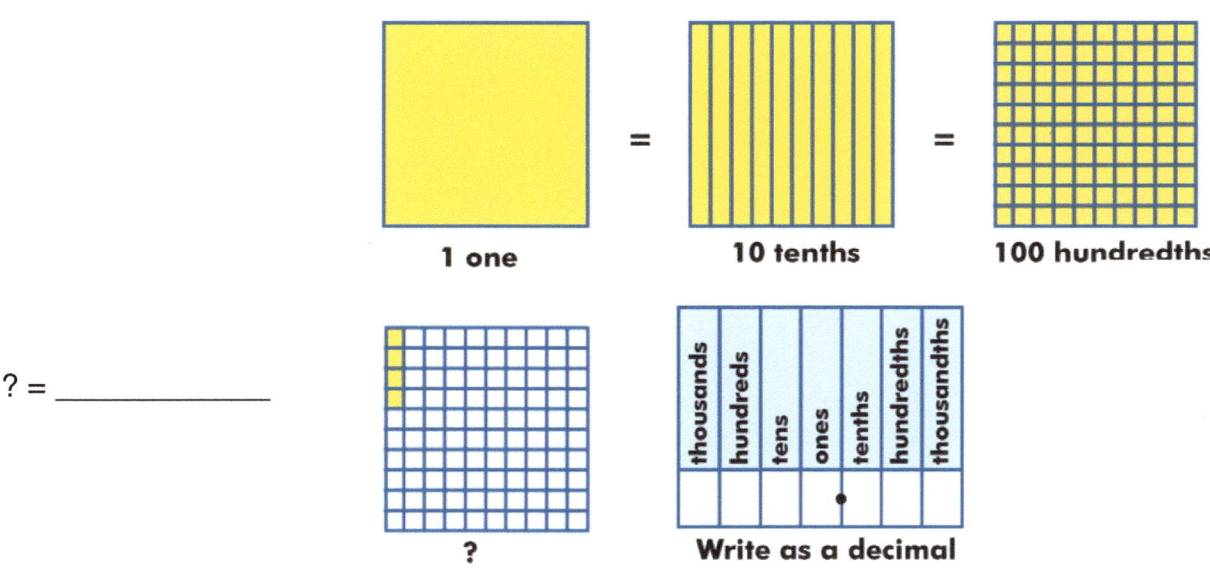

? = _____

Write the following as decimals.

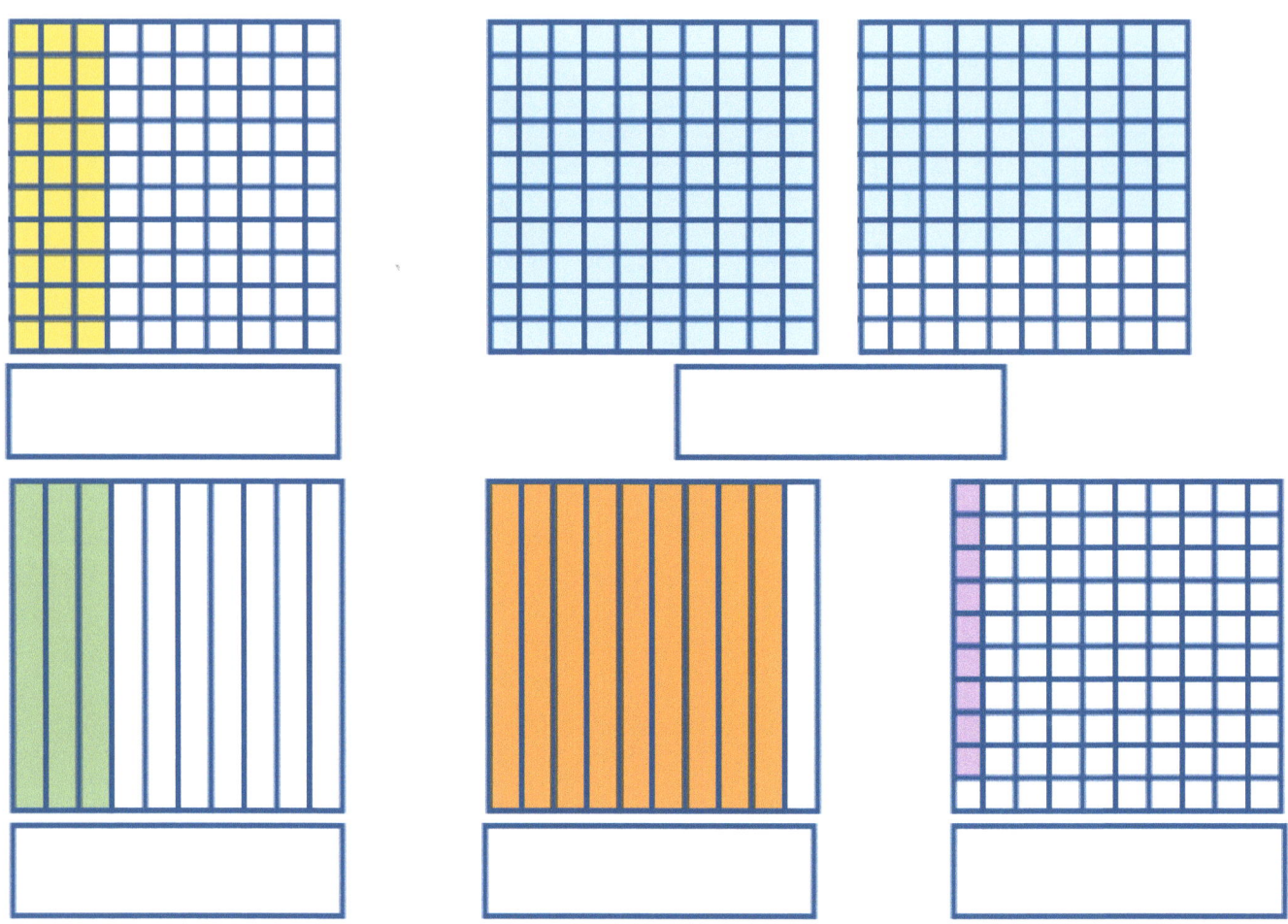

What does the 5 represent in each number?
No. 2 is completed for you.

0. 5 7

0. 0 4 5   0. 0 0 5

0. 8 5 1

Use the place card values to complete these decimals.
No. 1 is already finished for you.

(1) **Five hundred fifty-two thousandths** **(2)** **Seven tenths**

| 0. | 5 | 5 | 2 |

(3) **Forty-seven thousandths**   **(4)** **Three thousandths**

(5) **Nine hundred nine thousandths**   **(6)** **Forty-nine hundredths**

| 0. 0 0 2 | | 0. 9 | | 0. 0 0 9 |

| 0. 0 5 | 0. 4 | 0. 0 0 3 | 0. 7 |

| 0. 0 4 | 0. 0 9 | 0. 0 0 7 | 0. 5 |

Extra Credit

How would you write this as a decimal?

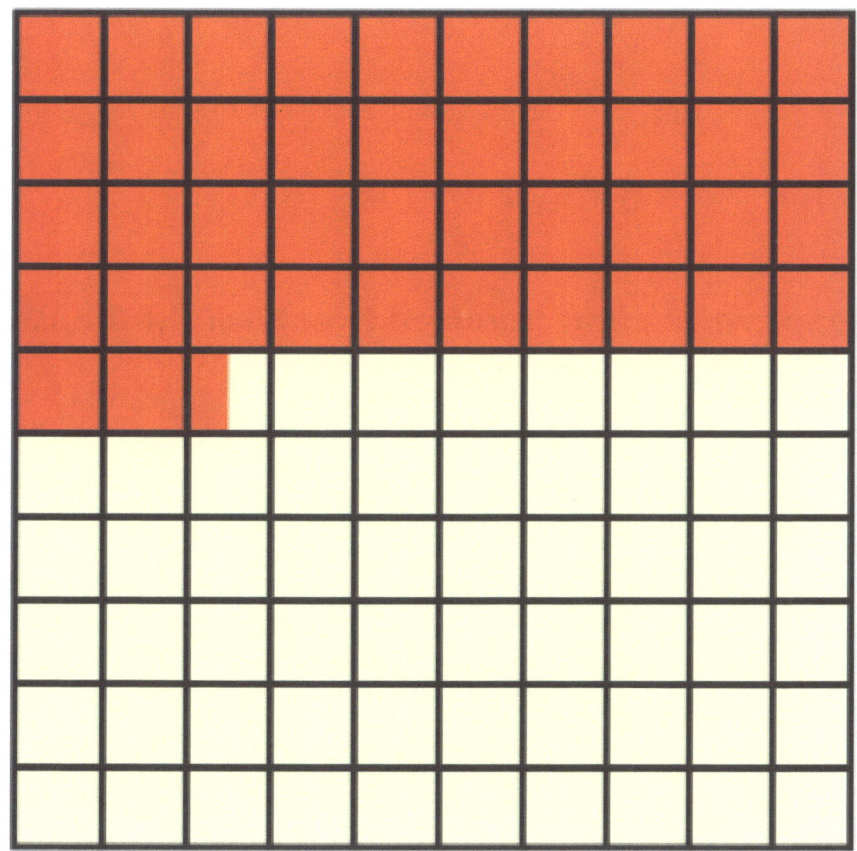

Name_____

Decimals Quiz

1 **True or false?   0.709 > 0.78**

2 **How do we write eight hundred twenty-eight thousandths?**

A 828,000

B 0.0828

C 0.828

D 0.80028

3 **What is the value of A?**

4 **What is the value of B?**

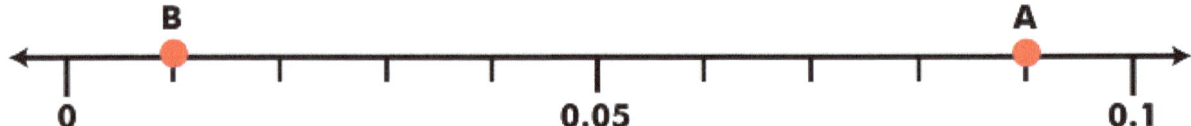

Add & Subtract Decimals

Key Vocabulary

hundredth

thousandth

rounding

Connect the one dollar partners.
Connect the kids whose change adds up to one dollar.

$0.24

$0.45

$0.61

$0.51

$0.55

$0.76

$0.39

$0.49

Adding Decimals.

Complete the problem below by modeling the answer and then adding the decimals.

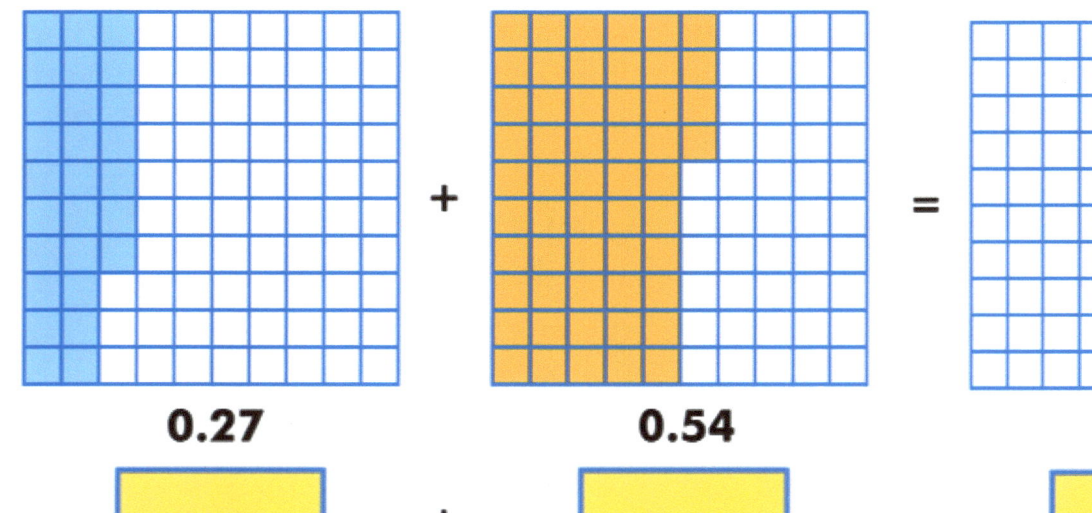

0.27   **0.54**

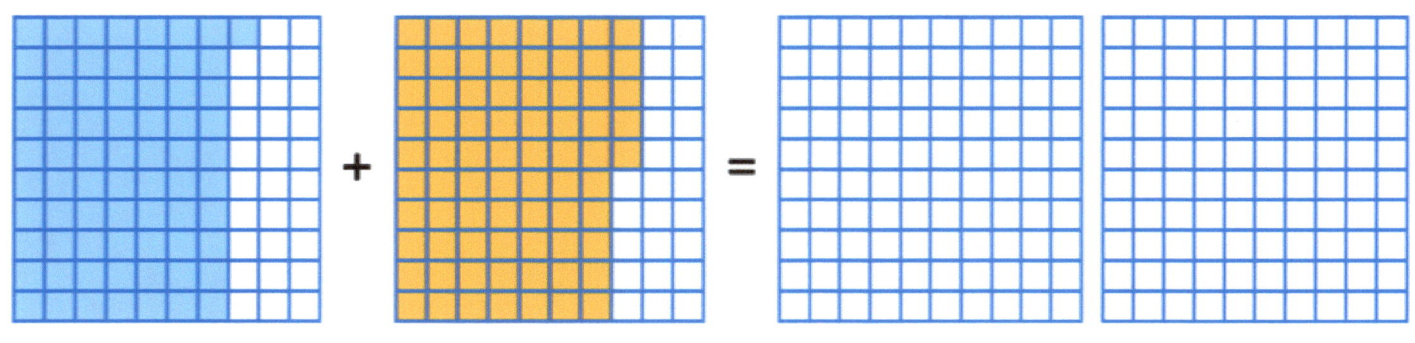

Subtracting Decimals.

Use the modeling to calculate your answer. Cross off squares to subtract.

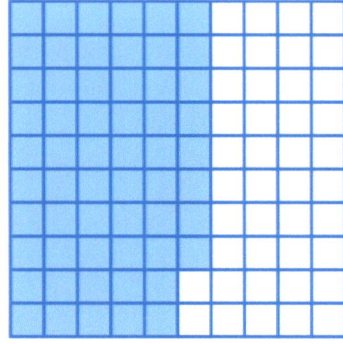

0.58 − 0.39 =

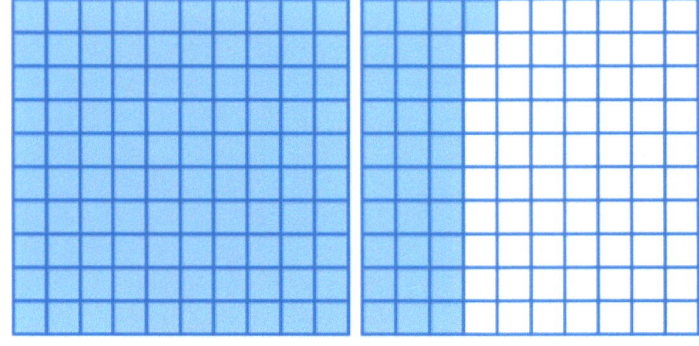

1.31 − 0.44 =

Adding Decimals

Study the problems below to discover how to add decimals. Complete the problem below.

| 3.5 + 4.7 | 7.9 + 5.074 |

$$
\begin{array}{r}
3.5 \\
+\ 4.7 \\
\hline
\end{array}
\qquad
\begin{array}{r}
7.9 \\
+\ 5.204 \\
\hline
\end{array}
$$

$$
\begin{array}{r}
\overset{1}{3}.5 \\
+\ 4.7 \\
\hline
8.2 \\
\hline
\end{array}
\qquad
\begin{array}{r}
\overset{1}{7}.900 \\
+\ 5.204 \\
\hline
13.104 \\
\hline
\end{array}
$$

4.465 + 2.54

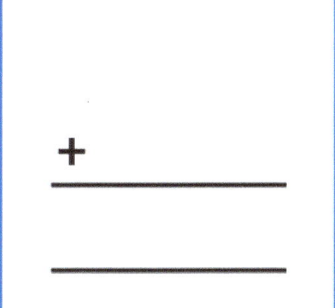

Subtracting Decimals

Study the problem below to discover how to subtract decimals. Complete the problem below.

$$2.7 - 0.56$$

$$\begin{array}{r} 2.7 \\ -\ 0.56 \\ \hline \\ \hline \end{array}$$

$$\begin{array}{r} 610 \\ 2.\cancel{7}\cancel{0} \\ -\ 0.56 \\ \hline 2.14 \end{array}$$

$$6.37 - 7.5$$

Name_____

Add & Subtract Decimals Quiz

(1) **What is the correct answer?**

$$16.897$$
$$+\quad 54$$
$$\overline{16.951}$$ ✗

(2) **0.5 + 0.5 + 0.5 = ?**

(A) 15

(B) 0.15

(C) 1.5

(D) 0.015

LUNCH MENU RECEIPT	
Chicken Fingers	$2.57
Garden Salad	$1.55
Soda	$0.99
TOTAL	

(3) **What is the total cost of lunch (in dollars)?**

(4) **How much change would you get from a $10 bill?**

Ratio & Proportion

Key Vocabulary

ratio

equivalent ratios

proportion

What is a ratio?

Ratio of boys to girls:

| 1:4 | 1 to 4 | $\frac{1}{4}$ |

A ratio of 1 to 4

Group of Students

Ratio of girls to boys: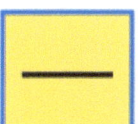

Ratio of boys to girls: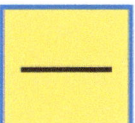

Ratio of girls to students: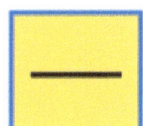

Complete these ratios.
If you aren't an artist use G for girl and B for boy.

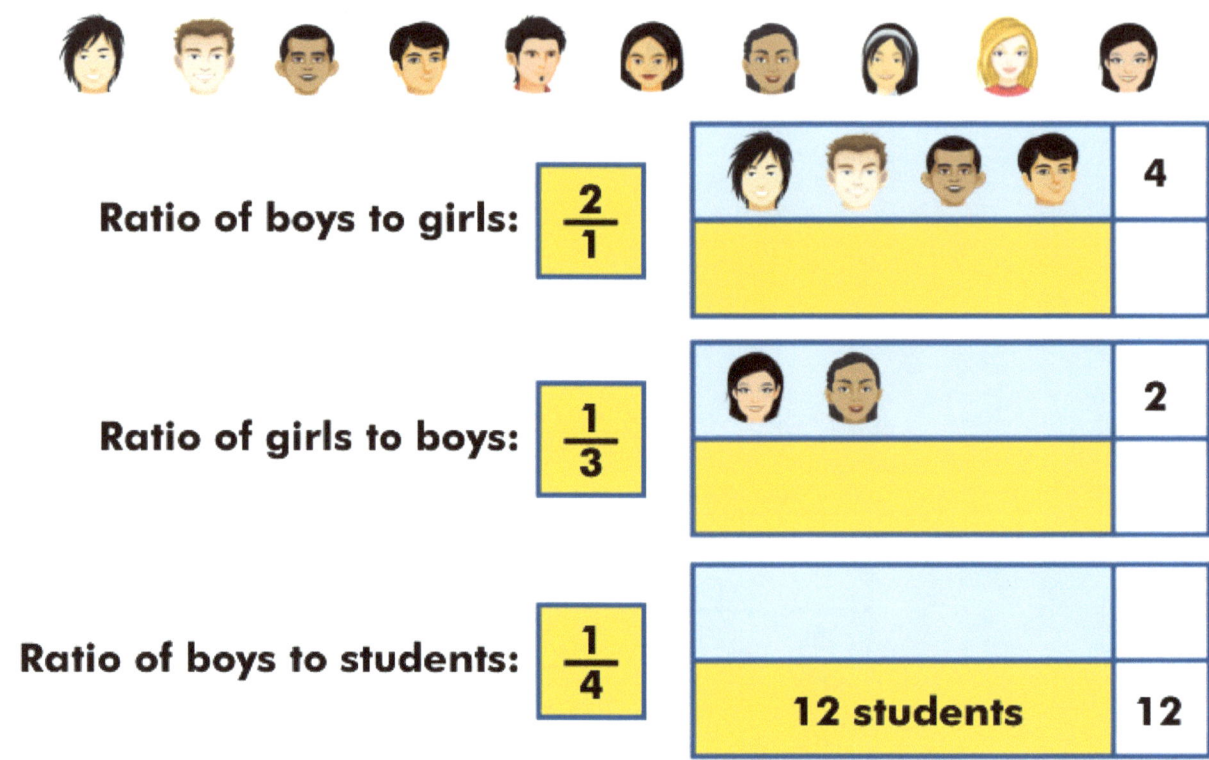

Ratio of boys to girls: $\dfrac{2}{1}$ 4

Ratio of girls to boys: $\dfrac{1}{3}$ 2

Ratio of boys to students: $\dfrac{1}{4}$ 12 students   12

Equivalent Ratios
Complete these equivalent ratios.

$$\frac{2}{1} = \frac{4}{2} = \frac{6}{} = \frac{}{8}$$

$$\frac{1}{3} = \frac{2}{6} = \frac{3}{} = \frac{}{27}$$

$$\frac{1}{4} = \frac{3}{12} = \frac{4}{} = \frac{}{24}$$

Solve problems using equivalent ratios.
Create the necklace by drawing the rubies and emeralds after you solve the problem.

**The ratio of rubies to emeralds on a necklace is 1:3.
If there are five rubies, how many emeralds are there?**

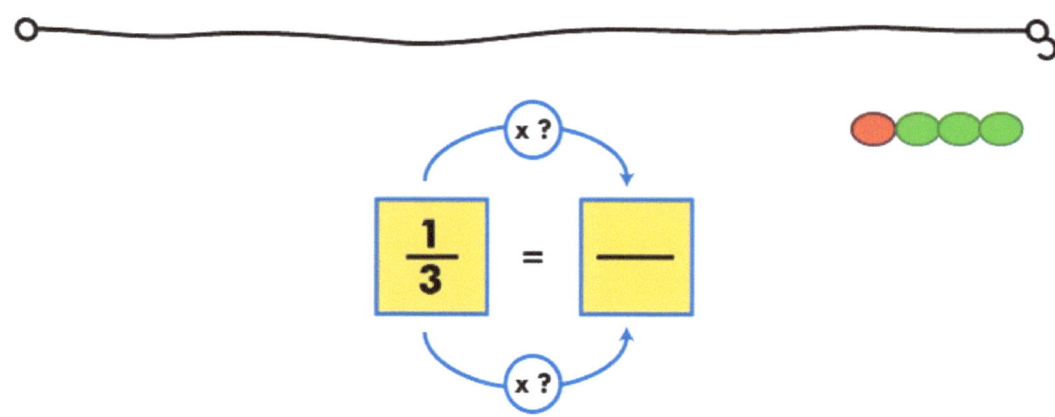

Use you knowledge

"Marinara Red" paint is made up of two parts yellow and five parts red.

The ratio of yellow to red is:

If 10 parts red are used, how many parts yellow are needed?

If 12 parts yellow are used, how many parts red are needed?

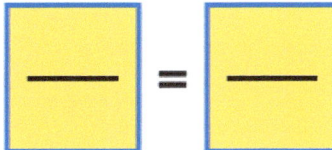

Name_____

Ratio & Proportion Quiz

1 True or false?
The ratio of girls to boys is 5:2.

2 Which ratio is equivalent to $\frac{10}{25}$?

A $\frac{1}{2}$

B $\frac{2}{5}$

C $\frac{20}{55}$

D $\frac{100}{255}$

3 If the ratio of boys to girls is 1:5 and there are 4 boys, how many girls are there?

4 If the ratio of girls to boys is 3:4 and there are 28 students, how many girls are there?

Percents

Key Vocabulary

percent

fractional percent

ratio

Percent means "per hundred."

25% of this grid is shaded blue

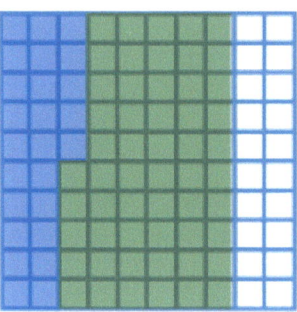

What is the value of the green shade?

What percent of the grid is not shaded?

Find the percents.
When possible write the percent as a decimal and a fraction.

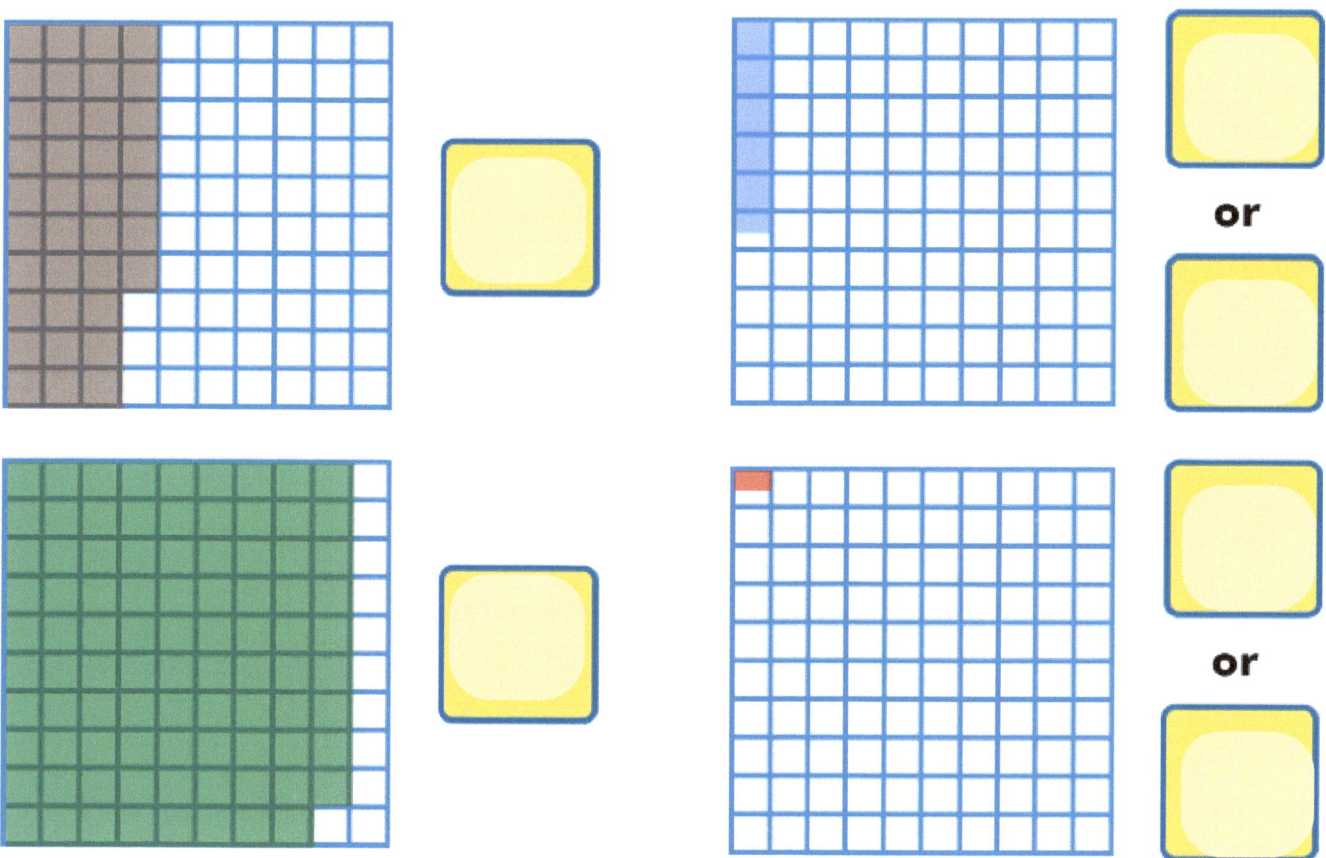

Practice percents with this vacation survey.

In a survey, 100 people were asked to name their ideal vacation. The results are shown in this 100 grid.

%

Ski

Cruise

Visit family

Beach resort

Theme park

TOTAL

What percent of the people surveyed chose each vacation?

Thanksgiving Lunch!
Complete the table.

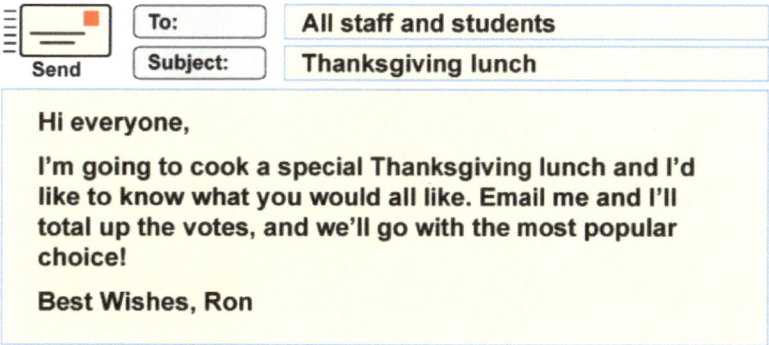

To: All staff and students

Subject: Thanksgiving lunch

Hi everyone,

I'm going to cook a special Thanksgiving lunch and I'd like to know what you would all like. Email me and I'll total up the votes, and we'll go with the most popular choice!

Best Wishes, Ron

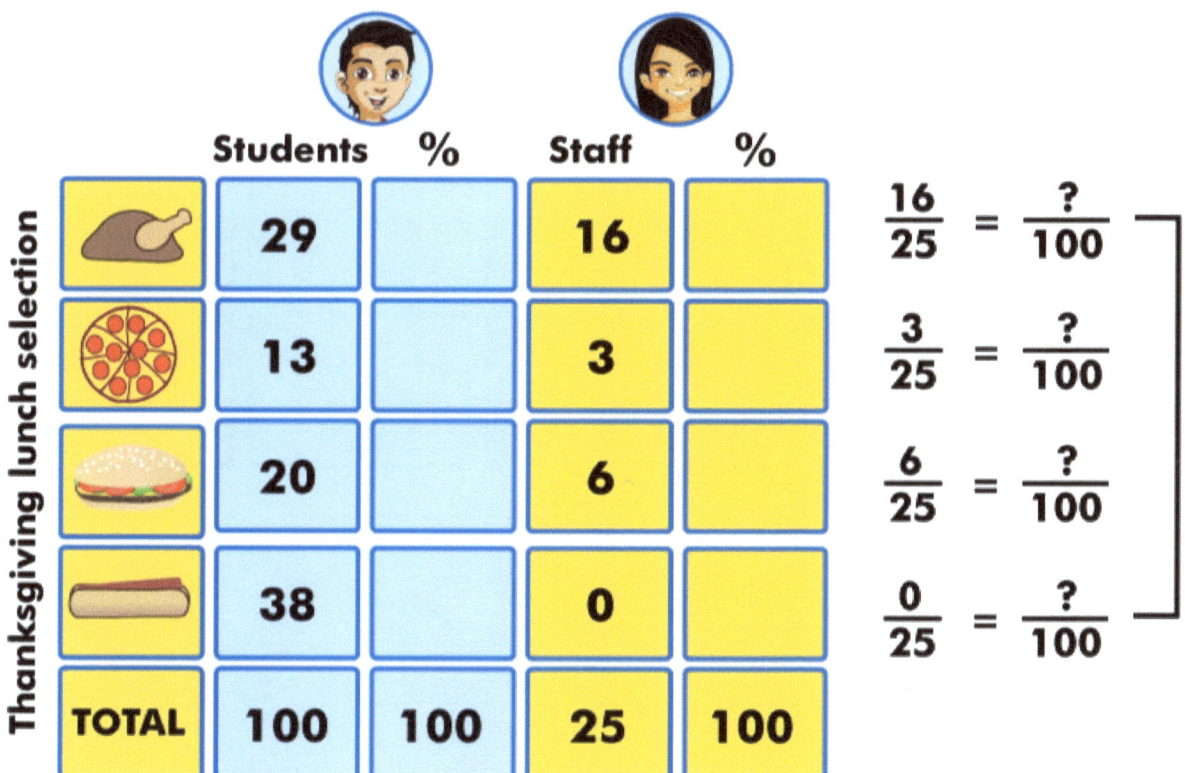

Thanksgiving lunch selection

	Students	%	Staff	%
🍗	29		16	
🍕	13		3	
🍔	20		6	
🌭	38		0	
TOTAL	100	100	25	100

$$\frac{16}{25} = \frac{?}{100}$$

$$\frac{3}{25} = \frac{?}{100}$$

$$\frac{6}{25} = \frac{?}{100}$$

$$\frac{0}{25} = \frac{?}{100}$$

Find the missing values.

Survey Results	Ratio	Out of 100	Percent
3 out of 10 people had visited Europe.	____	____	
24 out of 25 trains ran late yesterday.	____	____	
4 out of 5 cars in the parking lot were SUVs.	____	____	
15 out of 20 students said they liked math.	____	____	

What is Mayor Lewis' approval rating?
Another method for finding percent from a ratio.

 33 out of 40 people interviewed gave Mayor Lewis a positive approval rating. Express Mayor Lewis' approval rating as a percent.

1. **Write as ratio** $\dfrac{33}{40}$

2. **Divide numerator by denominator** $\quad 40\overline{)33.000}$ with quotient 0.825

3. **Multiply by 100%** $\quad 0.825 \times 100\% = 82.5\%$

More about Mayor Lewis.

See if you can figure out how many people approve of the Mayor's education policies. If they approve of the policy, they will vote for him in the next election.

 **Only 23 out of 40 people surveyed, approved of Mayor Lewis' education policies.**

What percent approved of the Mayor's education policies?

What percent disapproved?*

What percent will vote for the mayor at the next election?

*****Figures assume those who didn't approve disapproved.**

Name_____

Percent Quiz

1 True or false? 94.5% of this grid is not shaded.

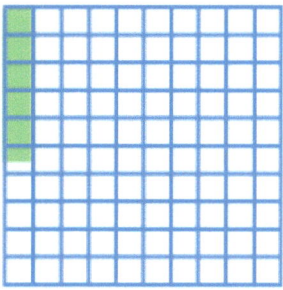

2 In a class of 25 students, 8 have a cell phone. Write this as a percent.

8%	16%	25%	32%
A	**B**	**C**	**D**

3 In a class of 25 students, 13 travel to school by bus. Write this as a percent (number only).

4 In a group of 40 athletes, 27 are female. Write this as a percent (number only).

Newburyport, MA 01950

1-800-596-3175

OnBoard Academics employs teachers to make lessons for teachers! We create and publish a wide range of aligned lessons in math, science and ELA for use on most EdTech devices including whiteboard, tablets, computers and pdfs for printing.

All of our lessons are aligned to the common core, the Next Generation Science Standards and all state standards.

If you like our products please visit our website for information on individual lessons, teachers licenses, building licenses, district licenses and subscriptions.

Thank you for using OnBoard Academic products.